"Patriarchy conflates women with nature while erasing their names from the history of science. *Her Wilderness Will Be Her Manners* corrects patriarchy on both counts, naming and celebrating women naturalists who worked in the field and in museums. Mangold brings these adventurous and resourceful women out of the acknowledgment pages of others' books and into the foreground of her own, highlighting in the process how science, in the guise of objectivity, dresses the natural world in conspicuous artifice. Using verbal and visual collage to evoke and trouble the tropes of collecting, preserving, classifying, and displaying specimens, Mangold fashions out of a trove of found fragments a fabulous feminist *Wunderkammer*, 'proof of woman's work' that does not disavow 'its feminine identity.'"

—Brian Teare, author of *Doomstead Days*

"Sarah Mangold's poetry of preservation is kin to Susan Howe's archival work. It is both haunted and haunting."

—Rae Armantrout, author of *Conjure*

Her Wilderness Will Be Her Manners

POETS OUT LOUD

Elisabeth Frost, *series editor*

Her Wilderness Will Be Her Manners

Sarah Mangold

Fordham University Press   New York   2021

Copyright © 2021 Fordham University Press

All rights reserved. No part of this publication may be reproduced, stored in a retrieval system, or transmitted in any form or by any means—electronic, mechanical, photocopy, recording, or any other—except for brief quotations in printed reviews, without the prior permission of the publisher.

Fordham University Press has no responsibility for the persistence or accuracy of URLs for external or third-party Internet websites referred to in this publication and does not guarantee that any content on such websites is, or will remain, accurate or appropriate.

Fordham University Press also publishes its books in a variety of electronic formats. Some content that appears in print may not be available in electronic books.

Visit us online at www.fordhampress.com.

Library of Congress Cataloging-in-Publication Data

Names: Mangold, Sarah, author.
Title: Her wilderness will be her manners / Sarah Mangold.
Description: First edition. | New York : Fordham University
   Press, 2021. | Series: Poets out loud
Identifiers: LCCN 2021016628 | ISBN 9780823297702 (trade
   paperback)
Subjects: LCGFT: Poetry.
Classification: LCC PS3563.A473 H47 2021 | DDC
   811/.6—dc23
LC record available at https://lccn.loc.gov/2021016628

Printed in the United States of America

22 21 20   5 4 3 2 1

First edition

Contents

Foreword vii

Her Wilderness Will Be Her Manners 1

Notes & 81

Foreword: Her Wilderness Is Her Language
Cynthia Hogue

> What interested me was
>   the way ladies survive
> as acknowledgments
>   in other people's prefaces
>   —Sarah Mangold

It was a privilege to read more than twenty worthy manuscripts, from which I selected, with no small amount of deliberation, *Her Wilderness Will Be Her Manners* for the Poets Out Loud Prize. Reading the lines above, drawn from Sarah Mangold's austere, book-length meditation on gender, art, and natural history, I am struck again by what compels me about this collection. The volume radiates with incisive insight. The power of its "arrested moments," the collages of thought and image in this lyric long poem, is cumulative. Here and there among passages of careful erasure and resonant assemblage are direct statements, scraps of fact that Mangold has mined from her material, which leap out at a reader, fresh and startling. She spent years researching natural history texts, seeking traces of early women naturalists whose record of works had been erased. Since historians do not "see women," as Mangold observes at one point, we must ourselves seek what has been overlooked, in order to comprehend how gender affects perspectives on nature and history.

    The serial poems that comprise the major part of this remarkable collection (punctuated with visual collages) do not make whole an erased history. Nor do they imaginatively resurrect a broader range of forgotten women or re-create the lives of the women naturalists who are the volume's focus. Rather, Mangold "collected" the language of the poems by engaging in "conversations with books" (an act of "wreading" inspired by Charles Bernstein's notion of a merged reading and writing practice). Her own "errand into wilderness" was linguistic. She employed methods of poetic construction that, like the women naturalists she researched, are impeccably accurate to the fragmentary nature of her material. *Her Wilderness* questions the objectivity of inherited historical narratives by rendering perceptible the ideological bias of those accounts. To illustrate briefly, consider the assumptions—made apparent in the text's repetition—that condition the view that "women and plants alike flower momentarily," not only as they enter the realm of Symbolic representation, but also of Time. Mangold herself isn't impartial, of course; she is, like her poems, of necessity *partial*. These poems do not fill in the gaps, for in fact what is absent from the poems is as much the point as what is present.

Mangold investigates what she describes as "a different kind of truth" about women naturalists in the history of the natural sciences. She contemplates them as the scientists they were, making note of their rigorous desire for truthfulness and their "revolutionary" refusal to remain in the "feminine" realms of authority. She lists (without comment) the areas in which it was appropriate for white women of a certain class to "exert authority," such as the California Spring Blossom and Wild Flower Association and the Council for the Protection of Roadside Beauty. Mangold's engaged reading practice excavates facts tucked away in the "crevices of wilderness," the places to which these rebellious women were drawn. She finds signs of their labor—discarded notes and labels, out of which she makes poems and visual collages—and she names and describes them. Take, for dramatic example, Martha Maxwell, who harbored ambitions of "revolutionizing the world/ with regard to women" and who, concerned that the creatures she studied were vanishing, became a "strict vegetarian." She accorded these living beings an individuality, moreover, as she came to know them through her studies. Maxwell kept her "eyes fixed upon facts," however, because of the one truth she couldn't change:

"The world demands

proof of woman's work   I am unwilling
it should lose its feminine identity"

In meeting the world's demand, Maxwell took pains to be identified by gender, knowing full well that it was considered "extraordinary for a woman" to choose a life of intellect and "feminist action."

The challenge Mangold herself has so brilliantly met is how to employ the technique of erasure to write a work about an erased feminist history that might accurately reflect the dual actions of erasure and restoration. There is an ironic justice at the heart of *Her Wilderness Will Be Her Manners*, for the constructive methods of collage and assemblage can, through revision and strategic juxtaposition, clarify obscured content and reveal that which had been suppressed. The method of poetic erasure is not simply removing parts of an existing text in order to make new meaning but also discovering heretofore invisible meanings within time-worn narratives. "Daughters of time," as Mangold apostrophizes toward the volume's end, "become yourself." That lyric address gestures toward the place these vital women might yet hold. *Just so.*

Her Wilderness Will Be Her Manners

They put our body
into the text

and there we are
made to wonder

how far in
we have gone

Make us exclaim
in the space

of hissing
throat clearing

explicit instructions
how to look natural

      Famous angels

             diorama

                  intimacy from artifice

                     take a breath

                            peephole   nostalgia

                                        butterfly

                                              on a dime

                            arrested moments of social relations

                            capitalize the landscape

Who cannot sympathize
    with her enthusiasm

        all diseases arising
  from the shadow

 are incurable
    it rushes among them

A particular person
    keen capturing soul

      the describer must
 be remembered

 to my shirt is pinned
    badge of femininity

I remain a mere
   photographic plate

     going ashore gladly
classifier for her own reception

Scores of tiny throats
   of unconscious praise

     philosophic naturalist
 true butterflies

 time moths
   I was interested in marking

    successive awakenings
           phenomena of a series

              of day-breaks as seen
    I was not the least

    striking part of the scene
           With palpitating heart

sugared trees
      lantern light

           splendid son
    lovely bride

  by the setting sun
      a description

before admiration
      animal life habitually exists

        in these awful solitudes
  breath barren

  crevices of wilderness
      I have often been charmed

of other occasions
      whirring wings

      woodland parks
    described as disposition

"It may be true that landscape painting tends to naturalize ideology." Taking my eye off the water cask and fixing it on the scenery where I meant it to be. Saying firmly in pencil in margins. "Help I am drowning."

When I botanize
I am thinking
When word and object coincide
Words are the shadow

Flicker   fox   flotilla
flowers and boot oil
beady eye   Dear devotion
river otter   red fox
ranged evergreen

I was by nature averse to classification and counting.
Women and plants alike flower momentarily entering

realms of history and speech. A battery of hidden
fluorescent tubes between daylight and moonlight.

Female endeavor and scientific classifiers. Promising
nothing less than the reproduction of space.

Unblind ourselves
from such

sun-worshipping
expenditures

"Jane Goodall and her mother made
2000 spam sandwiches for

fleeing Belgians before
embarking for the wilds of Tanzania"

Names according to the time of their coming. She who delves among books in various dead and living languages. She who works

upon the skins of creatures. Foliage of such fragile character that a coyote might have some excuse for action. A case of parallel

development that occurs in methods as well as in nature. And there assembled. A broad and graphic presentation of the conditions.

A ghostly group of great ground sloths. The whole question of how far it is allowable to depart from actualities.

Ideas about where it was appropriate for women to exert authority: American Fuchsia Society, California Spring Blossom and Wild Flower Association, Business Men's Garden Club, Save the Redwoods League, Council for the Protection of Roadside Beauty

diagram for violet

diagram for buttercup

diagram for dandelion

diagram for daisy

diagram for pond lily

twine heavy and light

an evenly notched leaf

grooving the blade of grass

driving a pin down through

each foot into the soft pine bottom

What interested me was
        the way ladies survive
as acknowledgments
        in other people's prefaces
the way historians will not
        see women in the museum
unless she seeks them out
        a catalog of eighty thousand cards
on the family Gramineae

        a gentle knock of nerve

        this process of a false body

Mrs. Chapman found it physically
impossible to bring away the water
soaked nests of the flamingos   Miss
Cherrie found equal difficulty with
the sodden nests of the guacharo birds

Miss Ruth Billard took notes
on painting methods
how to paint shadows
what paint was on her palette
how to paint skies
how to draw receding perspective
how to size birds on the painted background

No one could be sure which observation would prove

useful   celestial winds

                rainbows

                kidneys   gesture

of remembrance   perishing the keeper

        footless birds

              of paradise

Shall we fabricate our soil   Fake our trees   There
        are different kinds of truth   Brought to great

perfection   Here by Miss Mary C. Dickerson
        A single mold making many individuals   Cast in

gelatin   Bent into diverse attitudes   Are evergreen
        rammed full of straw

How delicately they colored
        no one could conceive

        Modeling tongues
casting by amateurs

at an ordinary fire
        madder brown

        madder lake
eyelids managed

Pounded white sugar
        powdered upon foliage

        Should the sea be close at hand
swarm in some shallow waters

Mounted on wires
        whipped with white silk

To record subtle changes in sky color
A grid of permanence
Ultramarine Blue
Cobalt Blue
Winsor Blue
Cadmium Yellow Pale
Cadmium Yellow Deep
Yellow Ochre
Indian Red
Cadmium Scarlet
Alizarin Crimson
Permalba her white

Everyone seems to think
it extraordinary for a woman

some nicety will be required
in painting betwixt hairs

considerable nicety will be required
to give the specimen

the appearance of nature
all these colors are liable

to change immediately after death
reddish brown fold

touched by scarlet varnish
powdered burnt umber

all the varnish colors
have a tendency to shine

my historian's notes
and labels disposed

Miss Alice Eastwood kindly
> put me in possession

of distance   of airiness   of boughs
> > of evergreen

Outfits and hints of preparation
> gestures of remembrance

plead the name be stabilized
> between   beside

captured & cured
> hang their harps on willows

The failure of the flamingo group
> their honesty of purpose

only angels can do without skin
> we are at liberty

to vary attitudes
> put as little life

as possible in the corners
> let some room to spare

My own chosen world
of intellectual development

and feminist action
might indeed unstring

unnerve
and unfit me

I add to her distribution
with imitation as an act of imagination

To vibrate like air in the distance
between the back wall and the spectators glass

The lakeshore observers
lighting forest vegetation

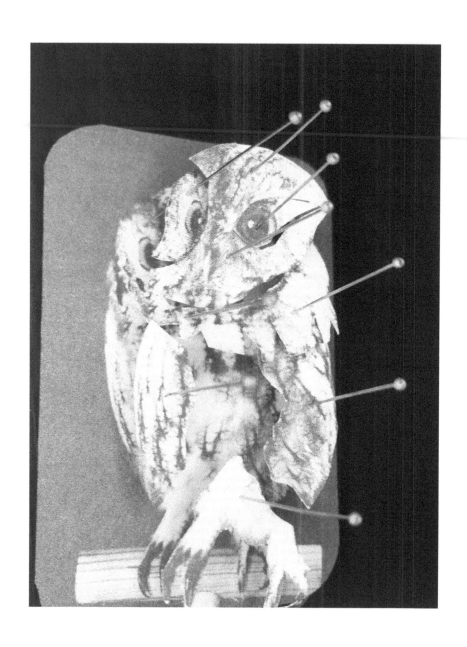

But when she moved in other circles she would choose to control certain tendencies. What are we to call ourselves? A peculiar sort of animal only capable of seeing a certain set of things. Drops of black

sealing wax to represent eyes she had removed. Neither had a surplus of elegant leisure. We would recommend you not to be deterred. Bring any object from the circumstances. Preserve every

natural curiosity you may meet. There was already so much to say. Butterflies must be put into your pocket-book. Beetles may be Drowned altogether. Soaked in water, cleaned by ants.

Let us dissect our feelings
Trees—How to understand them
Clouds—How they float
The Extraordinary and Bizarre
Painting from Memory

I can recall but few landscapes
the feeling that clouds
are always coming
within their short careers

Let us consider the prime difference in feeling
if you note the general
grasp the particular
Let clouds float   Render their edges

What everyone feels for nature   suppose those
feelings   An index of her sensitivity

Accommodations of translation
long understood as elegy   We feel

and pronounce this sunset a kind of microphone

Blown all over with pinholes or other holes
Skin is afterward   There is no discrete I

Each bit of nature landed in one column or
the other   Skin beamed down for the apparent

bending   Iridescent peacock feathers   Wings
of certain dragonflies

To the possibility that species
might be linked with gender
their individual deaths
wax in which to embody her ideas
foliage arranged according
to the season to be represented
my outfit came as near perfection

Women occupied many
kinds of places I do
not think of landscape
as without an unconscious

You must know how Nature
found me and I found Nature

Give the animal its exact attitude
then proceed   A movable horizon

Your hand will nearly always
be able to keep up

I am by nature a heavy percentage of embodied spirits. Like all specimens of any kind carefully ticketed as to its giver and its district. Your ship as I tell you. The lonely impulse of delight. How did the walrus fare? She accepted its existence, though uncertain in what class to place it. Emphasis on excluding not only the fabulous but also the uncertain. I described as if the manuscript descriptions were still in order.

Do not force
all the stuffing
a skin will hold

it is wives
who clean up
the work room

Fertility of couples interlock
collections of arguments
rationalize the randomness of fatality

to have a body
conditions of existence
or probabilities of life

Of many women's
invisible careers

to animals as they
are supposed to be

set-aside confidence
in one's own body

With all evidence of attachment obliterated
she worked warm fingers
placing flippers on seals
and legs on penguins
how cautious one must be
with the dry display
of free-swimming animals

"This sounds discouraging to a person whose occupation necessitates going about considerably in boats. My continual desire for hair-pins and other pins. My intolerable habit of getting into water. Abominations full of ants."

But this is becoming a feminine chapter

      Romantic by right of love   appropriation and appreciation

      She described her errand into wilderness

in language   how to ride, how to dress for it, how to shoot,

How to woman-who-goes-hunting-with-her-husband

      by the possessive voice a western abbreviation

           in great favor   typical affectations

of the hunter hero autobiography

   I am still a woman and may be tender   Hundred dangers which

seemed made to annihilate me

DELIA J. AKELEY
Jungle Portraits

MABEL MAXWELL BRACE &
MARTHA MAXWELL
Thanks to Abigail,
a Family Chronicle

MAXINE BENSON
Martha Maxwell, Rocky
Mountain Naturalist

MARY HENRIETTA
KINGSLEY
West African Studies

ALISON BLUNT
Travel, Gender, and
Imperialism: Mary
Kingsley and West
Africa

MARY DARTT THOMPSON
On the Plains, and
among the Peaks, Or,
How Mrs. Maxwell Made
Her Natural History
Collection

"Woman's Work"

Plunged you down plunged
down by one minor detail
I would have liked to
imitate a certain kind of
writing "Quotation marks
verify the existence of the words
in another reality"
Her own x-ray like vision
furs my landscape

Side view of a tiger's tongue

End view of a tiger's tongue

Top view of a tiger's tongue

On the appearance of the animal without its skin

she braced against the inequalities of the bark and drew

herself up among branches   The giving of names

to individuals involves an act of will

                    Wife, Mother, Householder

                    Film, Flashlight, Telephoto lens

She holds a well-accredited and happy record

Martha Maxwell had no leisure to be oppressed by loneliness
This desirable object attained by story

"To know there was at least one lady
who could do such a thing

Embraced several specimens
of high scientific interest

I regard it as one of the most valuable
single collections I have ever seen"

Legs holding electric light imprint nature as tamed
        Martha Maxwell's desire
               for truthfulness in art

knowing every animal intimately
        as personal acquaintance
            portrait of a friend   transmission
of books and bodies

between west & east
                digest & catalog
        reformist and strict vegetarian
        vanishing species

eternally secure in their true likenesses
            claiming the animal as a scientific specimen
        "The revolutionizing of the world
            with regard to women

is what I am fighting for—
        working for—and determined
to help accomplish"

Attribute the practice to desire
articulation of the visible proof of experience
skin disrupts the distance
the complete record of frontier
passing into history
devotion to the task at hand

Practice such expertise
variant of the skin trade
such doubts notwithstanding
as I pause in the midst of this recital
Dear
        Evidence of her desire
an artifact of its takeover

To send all the skins she wanted
suggestions of stealthy theft
masculinity persistent as a text
      rosy finch heart &c. &c.
I kept hold of my departing senses
doubt cast and describer suspected

so much to be remembered by hereafter

around a pond rendered in glass and wax

seasonally incongruous fur phases

Between cascade and lakelet appear irregular
vine-fringed mouths   She must carry the capacity

to be read   Keep eyes fixed upon facts
Receive images as they are   "The world demands

proof of woman's work   I am unwilling
it should lose its feminine identity"

IN THE WORK ROOM.

Among the named forebears in her path to abstraction.
An unbroken succession of callers wearing the stuffed

bodies of our familiar songbirds. This intersection of
intention. I accordingly call to them. Into this

wilderness. Unto consideration. I select for illustration
unsuitable attire. Her peculiar habit. I wish to say

in living creatures I well remember. Question the
concept of anticipation. Give to them an individuality.

Loyalties which I have visited. She could not articulate
horizon. Deciphering divine off in the disaster.

Seeking what to do with sons. Born bird-lovers. For one who enters in person a score arrive by mail. We have become in truth species. I declare synonym. Formally before the Union and informally everywhere. Bird-Love proofreading kindred subjects. Conspicuous decoys. Only to see to become convinced. Facts forced to support belief. A line of flamingos and rosy sunset. Their extreme wariness. The correctness of your observations. Impress with importance of sentiment.

In the Department of Birds we welcome
the possible. Accuracy is more-or-less

proposition. Low down and close up.
An architect in need of work. During

the next five years Mrs. Morrill painted
African landscapes never seen. To create

a message for the future. To prepare
for her coming tasks.

A robin thrush or blackbird whichever you happen to have

    Skinning   Skin-making

        Master the robin

      Now be careful   Cut very slowly

At the eye and close to the head   Until you can see through

      Define the exact position of the eyeball

          Skin each wing

Stop the wrong-side-out process there

    You will readily see coaxing begin to crowd

          Wing appears   Wings into place

        Wings legs tail   Cross feet

  Conveniences of home and in a hurry

Hither come and hence departed   many a man
        to represent   birds in situations   somewhat

similar   Both a frame   and what a frame
        contains   Who is this we   Our difference

from trees   grasses   clouds   Whose nature
        is marked   Wilderness   Farm   City

Rocket from A to B a fit gymnasium for action O

consciousness Observable proof Alphabetized

Awake Apples and prescriptives Woman to forget

marital woe by fighting African Jungle Beasts

Landscape every specimen permanent fact

Enthusiasms for this fine thing and that

Mrs. Akeley Quiet and Gray
to Hunt Big Game & Study
Baby Life in African Wilds

Going Alone into Jungle on
Brooklyn Museum Commission
Carries Her Own Table Linen

As all the names
Delia Akeley
indelicate remarks

Of the inconsiderate
"His faith in the results
of my cooperation

Convinced and inspired
I agreed to forget Paris
for the time being

I went to the northern woods of Michigan
each group is an exact reproduction
of what I found on a given spot"

Photographs of forest
plaster casts of leaves
flowers and quantities of moss

Four years' manual labor
carried on her back
variety in each leaf

Seventeen thousand leaves
as an intentional desiring act

Method of Pinching a Butterfly

Manner of Skinning an Elephant

Delia Reiss Akeley (first wife)
Appearance of Bill buffalo
hunting divorces elephant
hunting illnesses marriage
breakup of marriage pet
monkey   interest in primates
see Kirinyaga expedition

Reluctant Pioneers and Gentle Tamers

Strenuous and Dangerous Forms of Recreation Appeal to Many Dianas

Again, the first thing to notice is an absence   Delia remarks
pointedly on insects weariness and failure   an ideal quarry

meeting her commitment   an invisible Dorothy Greene
transcribes Carl Akeley's biography from behind the divan

"The physical appearance of the book is itself an eloquent story"
to hold forever gaze of meeting   each organism a vital moment
                                        each a window onto

BARBARA T GATES
*Natural Eloquence:
Women Reinscribe
Science*

LESLIE MADSEN-BROOKS
"Challenging Science
as Usual: Women's
Participation in
American Natural
History Museum Work,
1870-1950"

SALLY GREGORY
KOHLSTEDT
"In from the
Periphery: American
Women in Science,
1830-1880"

ELIZABETH FAGG OLDS
*Women of the Four
Winds*

SHARON MACDONALD
*The Politics of
Display: Museums,
Science, Culture*

KATE SOPER
*What Is Nature? :
Culture, Politics,
and the Non-human*

"Women's Work"

Though all my bones collapse on the platform. How kind they were to me. How careful for my comfort. Women!

Ah! As I have recently differed unapproached by railway. Had not then the franchise better educated, better fitted,

well-too-do, cart to lecture. Antelopes and their native neighbors form an attraction, even among the many

fascinations of the century's gathered productions. On this journal how thoroughly I approve her conduct. Where she

has done on the one hand rash not marked. Record my view that it has been bravely done, and most usefully done.

Suggesting the altitudes at which they were found. The powerful, the soft, the splendid.

"To call ourselves outsiders
is a kind of lie

Sometimes we are forced into this location

sometimes we choose to inhabit it"

Normalized

and comfortable

We survive tangible
pinned paperbound

My blood was rehearsing a kind of speech
         irrational vibrations   trailing with frail young feet

Fatal and threatening
                                      social tendencies
      fatigue my investigations

Any landscape
      is the absorption and transformation of another

Blood runs all over

        this century grown from none

                Painted birds anywhere   over flaws

disembodied

      violet

        beaming knife

      ossuary

            This sensible experience of boundary

          vestige of landscape

To be mistress if only of a small tent

This century grown from none

    to ask how rare

        peculiar undisturbed

to my own hopes

the soft tubular hairs
filled with blood where
there is a break enough
will always remain
to mark the catastrophe

how to shape a heron skin

how to open a turtle

Into the concept of notation

Miss Agnes Chase rehearsed demeanor

a strange and almost startling appearance

relegated to the notes for other interested researchers

a beverage deemed suitable for a feminine throat

Where ladies do go

within the frontiers of femininity

articulating the consistent

obliteration of their activity

in what passes for history

these effects we call facts

fantasy over the scene

not the boundary

Outlined like a phone tree I was speechless O

memories O symbolic transgressions Migrate and

devour a strong-box full of meaning O unconscious

kingdoms Pioneering elucidation Poach their

intertwined abandonment Run it like a traffic light

faint with expectation

In some dioramas   mirrors were strategically

placed to reflect light into an animal's eyes   This

sensitivity   To stabilize from outside moments

Slightly more colorful than what is seeing in life

       Blue as snowshadow

       To make the card as blue

       As the shadow on the sheet

The romance of the group   On site at sunset

palettes leaned heavily toward purple

in certain shadows

              The Greater Koodo appeared

at the appropriate hour

              Blue reflected sky on foliage

              The trees in distance less purple

How fortunate it would be   if all invasions and

    conquests resembled those of these

            daughters of the skies

      Whose sight of trip to time of translation

        is not the person

           but a portrait misplaced

       Love signals

           to the travelers ears

beet sugar   snowy floor of winter diorama

I recommend the dictionary as a means
of getting further information

The flash of butterflies
When I use the word

I cannot even suggest names
I proceed altogether differently

I wished to read directly from the book of nature. A word is what it does. A lighthouse of irregular habits advertised as an exercise in immediacy. Her study of botany, monsters, and curiosities. Scientific tourists betray their creation as evidence. Spirits leaping from one embodied condition to another. Women equipment to thought.

Nature does travel outside

its place of emergence

Produce a patterned vision

of how to move

and what to fear

in the topography

"It is critique in

the deep sense that

things might be otherwise"

The colors mark different
        kinds of possibility

        fur different
kinds of intention

who shall say what
        the nature of that process was

        she carried a merry heart
the very motion of reasoning

let a heavy dot
        in place of noun

        likeness was built-in
an affair of affection

everything tells of happiness
        we cannot help but sympathize

Daughters of time   desire to hear
        sure accompaniments of the still

orange–tree cocoa–nut fern
        become yourself   contributor to science

wings thrown open like powdered gems
        it is hopeless to paint general effects

naming a multitude of objects
        such elements of scenery

a pan of paraffin and an old hairbrush
        create convincing drifts of snow

However the heroine   upheld her respectability   being
    located indoors   She sees nature working by herself

she sees   a shiver in July   Skylarking   our good ship
    makes but slow progress   I once placed in a sealed jar

I kept skins   lampblack   duly crumpling   setting
    artificial eyes   gluing hair   flesh   fawn   pine

The separate sorts of plants
stand out before your eyes

Observations on secret societies
the kindred subject of leopards

I was only a beetle
and fetish hunter

Burn all your notions
of sun-myths and worship

Gold-dust and rum
may formed and luxuriant

Golden surface

of tomorrow I task

with palette and turpentine

Scientists wanted painted

birds which were maps

Architectural expressions of aspirations

a bowl of milk for drawing

fat from a duck

"Breathing New Life
Into Stuffed Animals:
the Society of
American Taxidermists
1880-1905"

Stuffing Birds,
Pressing Plants,
Shaping Knowledge:
Natural History in
North America, 1730-
1860

VICTORIA CAIN
"The Craftsmanship
Aesthetic: Showing
Making at the
American Museum of
Natural History,
1910-1945"

ANN B SHTEIR
Figuring It Out:
Science, Gender, and
Visual Culture

RACHEL POLIQUIN
The Breathless Zoo:
Taxidermy and the
Cultures of Longing

KAREN WONDERS
Habitat Dioramas:
Illusions of
Wilderness in Museums
of Natural History

"Women's Work"

Notes &

The language in this poem was collected through conversations with books, and the processes of wreading, erasure, and collage. A list of source texts, infatuations, and inspirations is included here, with particular indebtedness to Rachel Polquin's *The Breathless Zoo*, found by chance at Powell's Books. The title *Her Wilderness Will Be Her Manners* is part of a sentence found in *Chateaubriand's Travels in America* by François-René de Chateaubriand. A nineteenth-century Frenchman's travelogue of a trip to America composed twenty years after the trip, based primarily on his readings and poetic imagination.

QUOTATIONS, IN ORDER OF APPEARANCE:

Charles Harrison, "It may be true that landscape painting . . . " ("The Effects of Landscape," in W. J. T. Mitchell, ed., *Landscape and Power*, 2nd ed. Chicago: University of Chicago Press, 2002).

Donna Haraway, "Jane Goodall and her mother made 2000 spam sandwiches . . . " (*The Haraway Reader*. New York: Routledge, 2004).

Mary Kingsley, "This sounds discouraging to a person . . . " (*Travels in West Africa*. London: Macmillan, 1897).

Donna Haraway, "Quotation marks verify the existence of the words in another reality" (*The Haraway Reader*. New York: Routledge, 2004).

Elliot Coues, "To know there was at least one lady who could do such a thing . . . " (American Association of Museums, *Proceedings of the American Association of Museums*, vols. 9-11. Pittsburgh, PA: The Association, 1908–17).

Martha Maxwell, "The revolutionizing of the world with regard to women is what I am fighting for . . . " & "The world demands proof of woman's work/ I am unwilling it should lose its feminine identity." (Benson, Maxine. *Martha Maxwell, Rocky Mountain Naturalist*. Lincoln: University of Nebraska Press, 1986).

Delia Akeley, "His faith in the results of my cooperation convinced and inspired" (Olds, Elizabeth Fagg. *Women of the Four Winds*. Boston: Houghton Mifflin, 1985).

Donna Haraway, "The physical appearance of the book is itself an eloquent story." ("Teddy Bear Patriarchy: Taxidermy in the Garden of Eden, New York City, 1908–1936." *Social Text*, no. 11 [1984]: 20–64).
Donna Haraway, "To call our self outsiders is a kinda lie . . . " (*The Haraway Reader*. New York: Routledge, 2003).
Donna Haraway, "It is critique in the deep sense that things might be otherwise." (*The Haraway Reader*. New York: Routledge, 2003).

Further sources of inspiration that remain in traces and fragments throughout the poem:

The Art of Taxidermy by John Rowley
Autobiography of a Bird-lover by Frank M. Chapman
The Breathless Zoo: Taxidermy and the Cultures of Longing by Rachel Poliquin
Carlson's Guide to Landscape Painting by John F. Carlson
Ecology without Nature: Rethinking Environmental Aesthetics by Timothy Morton
"Goethe's Botany: Lessons of a Feminine Science" by Lisbet Koerner
Habitat Dioramas: Illusions of Wilderness in Museums of Natural History by Karen Wonders
Landscape and Power by W. J. T. Mitchell
On Her Own Terms: Annie Montague Alexander and the Rise of Science in the American West by Barbara R. Stein
"Painting Actuality: Diorama Art of James Perry Wilson" by Michael Anderson
Museum Politics: Power Plays at the Exhibition by Timothy W. Luke
Practice of Everyday Life by Michel de Certeau
Taxidermy and Zoological Collecting: A Complete Handbook for the Amateur Taxidermist, Collector, Osteologist, Museum-builder, Sportsman, and Traveller by William T. Hornaday and W. J. Holland
Vision and Difference by Griselda Pollock

The images in this poem were created in conjunction with writing and revising. Inspired by the collage process of my writing practice, I wanted to physically manipulate and inhabit the images that I was so intimately repeatedly viewing. I was also fascinated by the few surviving stereographic images of Martha Maxwell's taxidermy collections that are now often reproduced as a single one-dimensional image. Stereographs are made with a stereographic camera, which has two lenses and takes two simultaneous photographs roughly 2 1/2 inches apart. When printed and viewed with a stereoscope, the illusion of a single three-dimensional image is created. Using the tools of an insect collector, I cut my specimen images apart in an effort to preserve and display their uniqueness and re-create by hand the depth and perspective of the original image format. I also chose to preserve and display the "Woman's Work" of my research in bibliographic form.

Woman's Work, in order of appearance (source images in *Woman's Work* are in the public domain)

Mangold, Sarah. *Delia Akeley on Mount Kenya*. 2018. Riker mount, #2 insect pins, paper, foam board, acrylic paint, 8" x 6" x 2". Photographed by Lynn O C Thompson.

Mangold, Sarah. *Otus Aso Maxwellia (Martha's owl)*. 2018. Riker mount, #2 insect pins, paper, foam board, acrylic paint, 6" x 5" x 1 1/4". Photographed by Sarah Mangold.

Mangold, Sarah. *Woman's Work Bibliography #1*. 2018. Riker mount, #2 insect pins, paper, foam board, acrylic paint, 8" x 6" x 2". Photographed by Lynn O C Thompson.

> Delia J. Akeley, *Jungle Portraits*
> Maxine Benson, *Martha Maxwell, Rocky Mountain Naturalist*
> Alison Blunt, *Travel, Gender, and Imperialism: Mary Kingsley and West Africa*
> Mabel Maxwell Brace & Martha Maxwell, *Thanks to Abigail, a Family Chronicle*
> Mary Henrietta Kingsley, *West African Studies*
> Mary Dartt Thompson, *On the Plains, and among the Peaks, Or, How Mrs. Maxwell Made Her Natural History Collection*

Mangold, Sarah. *In the Work Room*. 2018. Riker mount, #2 insect pins, paper, foam board, acrylic paint, 6" x 8" x 2". Photographed by Lynn O C Thompson.

Mangold, Sarah. *Mabel Maxwell*. 2018. Riker mount, #2 insect pins, paper, foam board, acrylic paint, 6" x 5" x 1 1/4". Photographed by Lynn O C Thompson.

Mangold, Sarah. *Woman's Work Bibliography #2*. 2018. Riker mount, #2 insect pins, paper, foam board, acrylic paint, 8" x 6" x 2". Photographed by Lynn O C Thompson.

Leslie Madsen-Brooks, "Challenging Science as Usual: Women's Participation in American Natural History Museum Work, 1870–1950"
Elizabeth Fagg Olds, *Women of the Four Winds*
Barbara T. Gates, *Natural Eloquence: Women Reinscribe Science*
Sally Gregory Kohlstedt, "In from the Periphery: American Women in Science, 1830–1880"
Sharon Macdonald, *The Politics of Display: Museums, Science, Culture*
Kate Soper, *What Is Nature? Culture, Politics, and the Non-human*

Mangold, Sarah. *Woman's Work at the Centennial.* 2018. Riker mount, #2 insect pins, paper, foam board, acrylic paint, 5" x 6" x 1 1/4". Photographed by Lynn O C Thompson.

Mangold, Sarah. *Woman's Work Bibliography #3.* 2018. Riker mount, #2 insect pins, paper, foam board, acrylic paint, 8" x 6" x 2". Photographed by Lynn O C Thompson.

Mary Anne Andrei, "Breathing New Life into Stuffed Animals: The Society of American Taxidermists 1880–1885"
Victoria Cain, "The Craftsmanship Aesthetic: Showing Making at the American Museum of Natural History, 1910–1945"
Rachel Poliquin, *The Breathless Zoo: Taxidermy and the Cultures of Longing*
Sue Ann Prince, *Stuffing Birds, Pressing Plants, Shaping Knowledge: Natural History in North America, 1730–1860*
Ann B Shteir, *Figuring It Out: Science, Gender, and Visual Culture*
Karen Wonders, *Habitat Dioramas: Illusions of Wilderness in Museums of Natural History*

Thank You to the editors and publications where selections from this poem first appeared: *Anomaly, Bennington Review, Chaudilare, Conduit, Conjunctions, Electric Lit Commuter, Curious Specimens* (Sundress Publications), *Golden Handcuffs, h_ngm_n, Interim, jubilat, Kenyon Review, MiPOsiea, Ping-Pong, Poor Claudia, Sand, Shake the Tree Volume II* (Brightly Press), *Talking about strawberries all of the time,* and *Versal.*

Thanks also to rob mclennan of above/ground press for publishing selections from this poem as the chapbook *Birds I Recall* and inviting me to read as part of the Ottawa International Writer's Festival in celebration of above/ground press's 25th anniversary. Selections also appeared in the chapbooks *A Copyist, an Astronomer, and a Calendar Expert* (above/ground press) and *The Goddess Can Be Recognized by Her Step* (dusie kollektiv).

Endless thanks and gratitude to the National Endowment for the Arts, Willapa Bay AIR, Artist Trust, Centrum and The Helen R. Whiteley Center at Friday Harbor Laboratories for your support of time and space between the Pacific and Salish Sea.

Thank you to Elisabeth Frost, Cynthia Hogue and Fordham University Press for selecting *Her Wilderness* for the POL Prize in the weeks before the pandemic.

& always, love to my family and husband, Paul, for your unwavering support and encouragement and to Don Mee Choi and Melanie Noel for the daily and cosmic inspiration, and especially to Heidi Broadhead, for her deep reading, listening, and editing as different versions of this book unfolded.

Sarah Mangold is the author of the poetry collections *Household Mechanics* (New Issues, selected by C. D. Wright for the New Issues Prize), *Electrical Theories of Femininity* (Black Radish Books), and *Giraffes of Devotion* (Kore). She is the recipient of a 2013 NEA Poetry Fellowship, as well as support from MacDowell, Djerassi Resident Artists Program, Willapa Bay AIR, Virginia Center for the Creative Arts, Whiteley Center at Friday Harbor Labs, Seattle Arts Commission and Artist Trust. She lives in Edmonds, Washington.

POETS OUT LOUD
Prize Winners

Stephanie Ellis Schlaifer
Well Waiting Room

Sarah Mangold
Her Wilderness Will Be Her Manners

José Felipe Alvergue
scenery: a lyric

S. Brook Corfman
My Daily Actions, or The Meteorites

Henk Rossouw
Xamissa

Julia Bouwsma
Midden

Gary Keenan
Rotary Devotion

Michael D. Snediker
The New York Editions

Gregory Mahrer

A Provisional Map of the Lost Continent

Nancy K. Pearson

The Whole by Contemplation of a Single Bone

Daneen Wardrop

Cyclorama

Terrence Chiusano

On Generation & Corruption

Sara Michas-Martin

Gray Matter

Peter Streckfus

Errings

Amy Sara Carroll

Fannie + Freddie: The Sentimentality of Post–9/11 Pornography

Nicolas Hundley

The Revolver in the Hive

Julie Choffel

The Hello Delay

Michelle Naka Pierce

Continuous Frieze Bordering Red

Leslie C. Chang
Things That No Longer Delight Me

Amy Catanzano
Multiversal

Darcie Dennigan
Corinna A-Maying the Apocalypse

Karin Gottshall
Crocus

Jean Gallagher
This Minute

Lee Robinson
Hearsay

Janet Kaplan
The Glazier's Country

Robert Thomas
Door to Door

Julie Sheehan
Thaw

Jennifer Clarvoe
Invisible Tender

Milton Keynes UK
Ingram Content Group UK Ltd.
UKHW021213290224
438644UK00005B/482